Basic Digital Electronics

Digital System Circuits and Functions

How They Work and How They Are Used

By
Alvis J. Evans

This book was developed and published by:
Master Publishing, Inc.
Chicago, Illinois

Edited by:
Gerald Luecke
Charles Battle

Printing by:
Arby Graphic Service, Inc.
Lincolnwood, Illinois

Photograph Credit:
All photographs that do not have a source identification are either courtesy of
Radio Shack, the author, or Master Publishing, Inc.

Table of Contents

Preface

In *Basic Electronics,* we explained the basic concepts and fundamentals of electronic devices and circuits. In this book, *Basic Digital Electronics,* we explain the basic concepts and fundamentals of digital system functions, and the circuits used to perform those functions and build digital electronic systems. No segment of electronics has expanded more rapidly than digital electronics, led by the small size, low power, high reliability and low cost of integrated circuits, where all the circuitry is formed and interconnected on a single chip of silicon.

Basic Digital Electronics explains in detail how digital circuits are designed and applied in systems that execute specific tasks. In a straightforward, easy-to-read language, using detailed illustrations, practical worked-out application examples solidify the understanding of the concepts as the discussions proceed. Anyone who wants to know about digital systems, and how engineers design digital circuits should be able to learn from the explanations in *Basic Digital Electronics.*

Basic Digital Electronics begins by identifying familiar systems to help understand the difference between analog and digital systems, and identifies the common functions required when designing digital systems. Digital circuits, and how they are designed, covering logic circuits that make decisions, sequential circuits for temporary storage, a variety of circuits to couple, convert and compute digitally, and memory circuits for more permanent storage of digital information are discussed in the next four chapters.

The wide application of digital electronics to systems would not be possible without integrated circuits. A complete chapter is devoted to explain the basic techniques and manufacturing steps used to fabricate, assemble and test integrated circuits.

The last two chapters conclude the learning by showing how the circuits and functions are combined into systems—first for computers, and then for digital communications.

Multiple-choice quizzes and a set of questions and problems at the end of each chapter reinforce the learning. Answers to quizzes are on the quiz page; worked-out answers to questions and problems are given in the Appendix.

Studying the material, understanding the worked-out examples, working the quizzes and solving the problems should result in an understanding of the basic concepts and fundamentals of digital electronics. That was the goal of writing *Basic Digital Electronics;* we hope we have succeeded.

A.J.E.
M.P.I.

CHAPTER 1

A Look at Familiar Systems

Comparing Analog and Digital

The field of electronics can be classified into two groups: analog and digital. Analog quantities vary at a continuous rate, whereas digital quantities vary in discrete values. We can find many examples of each all around us. All things that can be measured quantitatively in nature are either analog or digital. Let's look at some examples of both analog and digital quantities.

The difference between analog and digital can be seen at the entrance to your local library. There are the steps, which is the digital route; and there is the ramp, the analog route. In a wheelchair, you better take the analog route. Now consider pitching pennies at the steps and the ramp. Where would the pennies land? As we see in *Figure 1-1*, the pennies can land only on steps 1, 2, 3, or 4, whereas they can land anywhere at all on the ramp. The steps represent digital — a discrete group of values; the ramp represents analog — a continuous group of values.

Figure 1-1. Pitching pennies at a ramp (analog) and steps (digital).

Another example of analog and digital quantities is shown in *Figure 1-2*. Here we have two buckets—one filled with water and the other filled with marbles. Suppose that we want to measure the contents of each bucket. The procedure would be different for each. For the bucket of water, we could dip the water out with a measuring cup or pour the water into the measuring cup. In either case we would probably leave a little in the bottom of the bucket or perhaps spill a little. Our measurement would not be very precise. With the bucket of marbles we could simply count them as we removed them. Our measurement would be much more precise. And, if we repeated the process on each over and over again, we would most likely get a different value for the quantity of water each time. But the number of marbles that we counted would be the same each time. This degree of precision is a characteristic difference between analog and digital. Digital is more precise. And how about if you were asked to reproduce each bucket at a distant location. Suppose that you were told one bucket had 6⅔ cups of water, and the other bucket had 723 marbles in it. Which bucket could you reproduce more accurately? The digital one, of course, because you can count out a discrete number of units.

Figure 1-2. Two full buckets, one with water and one with marbles.

Familiar Examples of Analog and Digital Quantities

Most physical quantities that occur in nature are analog. Temperature can have an infinite number of values, even between 65° F and 66° F. Sound volume can vary continuously from very soft to extremely loud. The velocity and force with which a carpenter swings a hammer is analog in nature. If so many things in the world are analog, why do we use more and more digital to represent these things? The answer to this question is that electronic circuits can process information about these physical quantities easier and more accurately with digital than with analog.

Clocks

Many quantities are familiar in both analog and digital forms. *Figure 1-3a* shows examples of clocks. Modern mechanical clocks date from the late Middle Ages. Analog watches have served us well with improvements only in the basic design. For instance, by the end of the 15th century the spring had replaced the weight in some clocks, allowing them to be built small enough to be carried. Only within the last decade or two has the analog watch been almost completely replaced by the digital watch. (We wonder what the terms clockwise and counterclockwise will mean to the next generation.)

Thermometers

Figure 1-3b shows examples of thermometers. The analog thermometer has been used for temperature measurement since it was introduced around 1720 by Gabriel Daniel Fahrenheit (1686-1736). An example is a mercury thermometer, in which a rising or falling (expanding or contracting) column of mercury represents a rising or falling temperature. Digital thermometers have replaced these in most applications today, especially in the medical field.

Meters

Two basic types of meters are used to make electrical measurements. As shown in *Figure 1-3c,* the analog type has a needle that deflects along a scale and indicates the value of the quantity measured by the position of the needle on the scale. Measurements on a digital meter appear as a number on the display screen.

Audio Recordings

The fourth example of analog and digital is shown in *Figure 1-3d.* The human voice and musical instruments produce sound that is analog. The ear is also an analog device that responds to the sound signals. The first audio recording devices were analog. From about 1880 through 1980, the phonograph held the dominant position for audio recordings for the home. Around 1980, the audio cassette tape entered the scene. In the late 1980s came perhaps the most remarkable development in audio technology—the digitally recorded compact disc (CD).

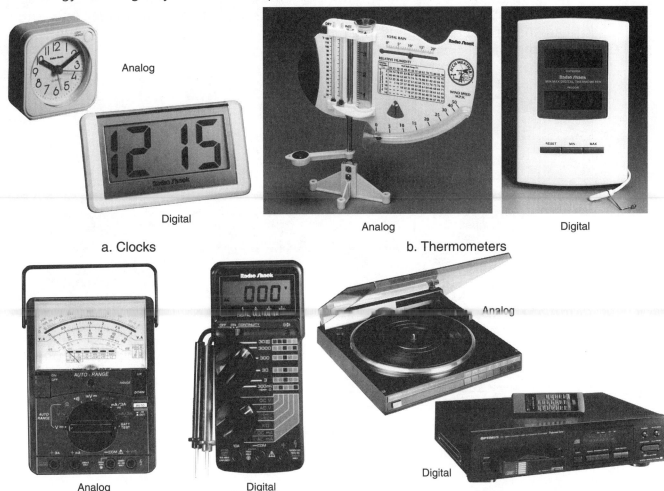

a. Clocks   b. Thermometers

c. Electrical Multitesters   d. Turntable and CD Player

Figure 1-3. Examples of the same quantities (time, temperature, electrical values, and sound) represented both as analog and digital values.

Recording devices for a CD convert analog signals to digital signals. The digital information is stored on the CD by making discrete indentations inline in a track on the surface of the CD. When the CD is played, a laser beam reflected from the surface converts these indentations into light pulses, which are detected and the digital data is converted back to analog signals by the CD player.

So why go to the trouble and extra steps to convert to and from the digital format? The CD digital audio vastly improves the audio quality over its analog counterparts—so much so that anyone who hears it will have a hard time going back to traditional analog audio. In addition, the digital format has virtually eliminated many problems of analog audio reproduction, such as noise and distortion due to fingerprints, smudges and scratches on the recording medium. Since a laser beam reads the information from the CD without mechanical contact with the CD, no matter how many times it is played, there is no wear on the disc.

Analyzing a Volume Control Circuit

Every radio and television receiver has a volume control, and most of them look like the circuit of *Figure 1-4a*. This is a partial block diagram of a radio receiver that shows the schematic of the volume control in more detail. The volume control is a continuously variable voltage divider; that is, it has an output voltage that changes as the control is varied. If the control is at the bottom (ground), there will be zero input voltage (V_{IN}) to the amplifier. If the control is at the top (maximum rotation), the input voltage to the amplifier is the full voltage, V_O. *Figure 1-4b* shows graphically the relationship between the input voltage and the amount the control is turned. Notice that the graph is a continuously varying line without jumps or breaks in it. The output is an analog of (or analogous to) the input voltage. The input voltage is a continuously proportional amount of the voltage V_O.

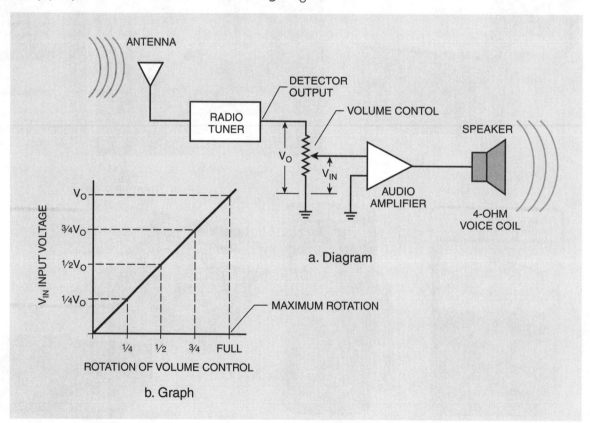

a. Diagram

b. Graph

Figure 1-4. Partial block diagram of radio receiver with schematic showing volume control in more detail. Graph shows relationship between V_{IN} and volume control rotation.

Example 1. Power Delivered as Volume Control is Varied

If the amplification (or gain) of the audio amplifier in *Figure 1-4a* is 100 and the signal voltage output of the radio detector (V_O) is 0.1 volt, determine how much power is delivered to the speaker with the volume control set at full volume, one-half volume, and one-tenth volume.

Use $P = \dfrac{E^2}{R}$, where P is the power in watts, E is the EMF in volts, and R is the resistance in ohms.

Full Volume — At full volume, the full 0.1 volt of the detector is applied by the volume control to the input of the audio amplifier. The gain of the amplifier is 100, so the speaker receives 0.1 × 100 or 10 volts. The power is $P = 10^2/4 = 25$ watts when the volume control is full blast. The resistance of the speaker is 4 ohms.

Half Volume — With the volume control at one-half volume, the audio amplifier input is 0.1 × 0.5 or 0.05 volt. The amplifier's gain of 100 delivers 0.05 × 100 = 5 volts to the speaker. This results in a power to the speaker of $P = 5^2/4 = 6.25$ watts. Notice that, though the voltage dropped by one-half, the power dropped by one-fourth.

One-tenth Volume — With the volume control at one-tenth of full volume, verify that the power to the speaker is 0.25 watt. The voltage is reduced by 1/10, but the power is reduced by 1/100!

The voltage varies in a linear manner and the power varies in a nonlinear manner, but both are analog. When we discuss integrated circuits (ICs) a little later, we will see them classified as linear and nonlinear. The term nonlinear in the case of ICs refers to digital. We must be careful in our use of the term nonlinear so it is clear what we mean.

What About a Race Track?

Is the speed control analog or digital for the miniature racetrack car shown in *Figure 1-5*? When the speed control slides from A to B, it causes the resistance of the rheostat to change linearly from 10 to 0 ohms. According to Ohm's law, the voltage and resistance of the circuit determine the current through the race car motor, which, in turn, controls the speed of the race car. The voltage is constant. The control resistance is at half-value when pressed half-way, and is at one-fourth value when pressed three-fourths way.

Example 2. Determining Current that Drives the Car Motor

What is the current in the circuit of *Figure 1-5b* if the speed control is set to one-half of the distance from A to B? The resistance of the car is 4 ohms and the resistance of the wire and track is negligibly small.

Use Ohm's law $I = \dfrac{E}{R}$, where I is the current in amperes, E is the EMF in volts, and R is the resistance in ohms.

$I = \dfrac{6}{(5 + 4)} = 0.667$ amperes or 667 milliamperes.

If the speed control is varied from A to B, the speed changes continuously as does the current. Therefore, this is an analog circuit.

The resistance varies linearly like the volume control of *Figure 1-4a;* however, there is a difference in the way the circuits operate. In *Figure 1-4a,* the control is in *parallel* with the detector output, and the rotating wiper selects different values of *voltage.* The *higher* the resistance value at the wiper, the louder the sound output. In *Figure 1-5,* the control is in *series* with the motor and power supply, and the rotating wiper varies *current* through the motor. The *lower* the resistance value at the wiper, the faster the race car goes.

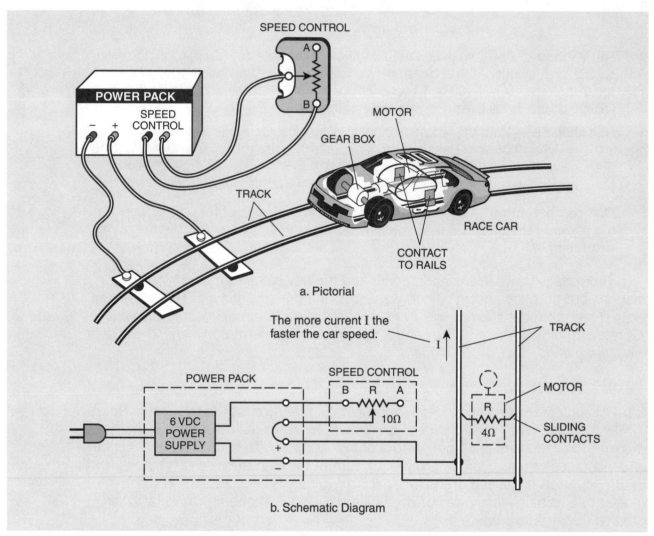

Figure 1-5. Speed controller for a miniature race track car.

A Digital Clock System

One digital system that is found in virtually every home is a digital clock. A simplified clock system that uses counters and frequency dividers is shown in *Figure 1-6.* These individual circuits will be discussed in detail in Chapter 4, but for now let's look at a overview of the clock system. The timing accuracy of the clock depends on the crystal oscillator, which does not vary with temperature, humidity, or power variations. Three counter blocks are shown in the block diagram: the first counts seconds, the next counts minutes, and the third counts hours. All counters output to the decoder which converts the count so it can be displayed. The maximum count of the first two counters is 59 (0 is a count, so there are 60 counts), the maximum count of the third is 12. The first and second counters start at zero, the third counter starts at one instead of zero so that it counts the hours from one o'clock to twelve o'clock. Each counter counts its input digital pulses and outputs a digital pulse to the next counter when it reaches its maximum count.

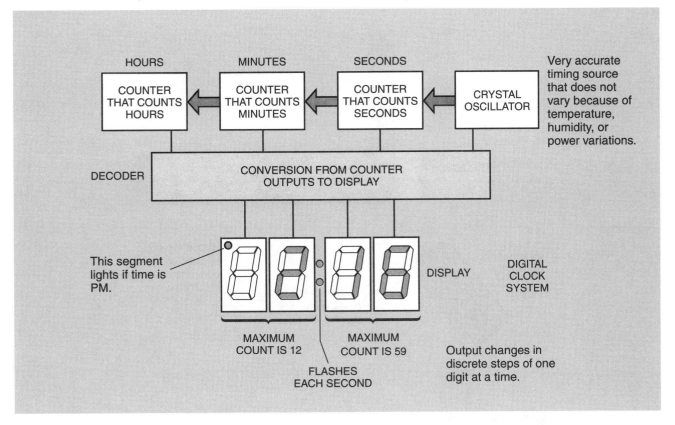

Figure 1-6. A digital clock system block diagram.

Example 3. Counting Digital Pulses

What time is it if the crystal oscillator in *Figure 1-6* has put out 37,200 one-second pulses. The clock is at 12:00 when the pulse counting starts, and with the first pulse the display shows 12:00:01.

Each hour contains 3600 one second counts, so 37,200 counts takes:

37,200 counts / 3600 counts / hr = 10.33333 hrs   (10 hrs)
3600 × 0.33333 = 1200 counts
1200 counts / 60 counts / min = 20 min   (20 min)

So the clock will show 10:20:00 after 37,200 pulse from the crystal oscillator.

We now have a good idea of analog (continuously varying) and digital (discrete parts) information. Since this book is about digital systems, from now on we will be dealing with digital information and digital systems. Let's look first at how information is represented digitally.

Codes

TV Remote Control

An example of using digital coding to represent information is as close as your TV remote control. A simplified version of the TV remote control system is shown in *Figure 1-7*. Each button on the keypad is assigned a particular code to cause a particular action in the TV. Pushing a button on the remote control generates a coded pattern that turns the infrared light transmitter on and off. The TV receiver detects the bursts of light and they are decoded into a required action, such as to turn on the power, change the channel, reduce or increase the volume, etc.

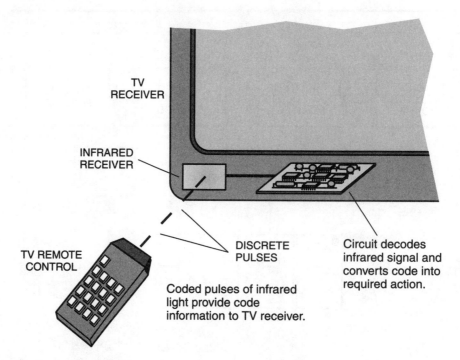

TV
RECEIVER

INFRARED
RECEIVER

TV REMOTE
CONTROL

DISCRETE
PULSES

Circuit decodes
infrared signal and
converts code into
required action.

Coded pulses of infrared
light provide code
information to TV receiver.

Figure 1-7. A simplified version of the TV remote control system.

Decoding Morse Code

Morse code is a signaling code first devised by Samuel F. B. Morse in 1838 for use with his electromagnetic telegraph. The code uses two basic symbols or signaling elements: the "dot," a short-duration electric current, which gave a quick deflection of the armature of Morse's receiver and so caused a dot to be printed on the strip of paper moving beneath the ink pen carried by the armature; and the "dash," a longer-duration signal that caused a dash to be printed. Using this code, the various alphanumeric characters needed to compose a message (letters and numerals) could be represented by groups of these two signal elements. The International Morse Code is shown in *Figure 1-8* along with a circuit that can be used to send a message by Morse code from one place to another. In the International Morse Code, for example, one dot followed by one dash (• —) symbolizes the letter A; the sequence "dash dash dot dot dot" (— — •••) symbolizes the numeral 7. When transmitting, the dot and dash elements are separated by a time interval that has the duration of one dot; the dash has a duration equal to three dots. The space between characters, whether letters or numbers, is equal to three dots; the separation between words is equal to six dots. Morse's code rapidly gained acceptance and its importance is not just historical; it illustrates the simplicity of a complete digital data communications system. A two-state communications system is the simplest, the easiest to build, and the most reliable. The two states can be On and Off, or Light and Dark with light beams, or with two durations of a tone, or, any other system with only two possible values. A two-state or two-valued system is referred to as a *binary system.*

Number Systems

What Do Numbers Mean?

Numbers are the primary language of all digital equipment. The information processed by digital devices and systems, including computers, is usually numerical in nature. As we will see, even the alphabet and other symbols can be expressed as numerical codes.

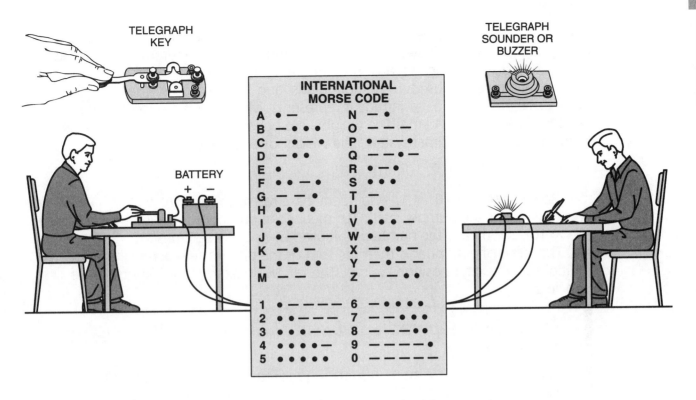

Figure 1-8. A circuit that can be used to send a message by Morse code.

Example 4. Morse Code Messages

Decode this message using the International Morse Code table in *Figure 1-8*. There are four words, one on each line.

Morse successfully demonstrated his magnetic telegraph by sending the above message in 1844. It said "What hath God wrought".

Now try this one. There are two words, one on each line.

Solution: Write the decoded message in the space below. The correct solution is given at the end of this chapter.

We usually think of numbers as their symbols, but the numerical symbols themselves are quite versatile. Their meaning or value can vary according to the way they are utilized. It surprises some that there are other number systems than our "everyday" system that we use for phone numbers, our check book, or our street address. These numbers that we use everyday make up what is called the *decimal number system*. Although decimal numbers are common, the weight structure is not always understood. Let's review the characteristics of the place-value structure of number systems.

Decimal Numbers

The decimal number system, with which everyone is familiar, uses the symbols 0,1,2,3,4,5,6,7,8, and 9. Because it has 10 symbols, it is also referred to as a base-10 system or radix-10, which means the same thing. If we wish to express a quantity greater than nine, we use two or more digits, and the position of the digit within the number determines the magnitude it represents. Most number systems use a positional weighting scheme. The weights are powers of the base that increase from right to left. In the decimal system, the weight of each position is

$$\ldots 10^5, 10^4, 10^3, 10^2, 10^1, 10^0.$$

Remember that $10^0 = 1$, in fact, *any base raised to the zero power is equal to one.* 10^1 means the number itself; 10^2 means the number multiplied by itself twice (10×10); 10^3 means the number multiplied by itself three times $(10 \times 10 \times 10)$; etc. The value of a number is then the sum of the digits after each digit has been multiplied by its digit position weight. The following example will illustrate the principle.

Example 5. Separating a Decimal Number by Its Weighted Digit Position

$328 = 3 \times 10^2 + 2 \times 10^1 + 8 \times 10^0 = [3 \times (10 \times 10)] + [2 \times (10)] + [8 \times 1]$
$328 = 300 + 20 + 8$
$328 = [3 \times 100] + [2 \times 10] + [8 \times 1]$

      ↑   ↑   ↑

  3rd digit 2nd digit 1st digit
  weighted weighted weighted
    value   value   value

Binary Numbers

Digital equipment uses a special number system to represent and process quantities. Because this system uses only two symbols or digits (0 and 1) to represent the quantities, it is called the binary number system. Because of the inherent bistable nature (having two states) of electronic circuits, binary numbers are more easily and quickly processed by electronic circuits than other numbers. Therefore, virtually all digital equipment uses binary numbers.

The binary system has two digits, 0 and 1, called **bi**nary digi**ts** or bits. It is a base-2 system, just as the decimal system with its ten digits is a base-10 system. To count in the binary system, we start with 0, count to 1, and we have run out of digits. So we can count only two values. We must go to the next digit position and continue. It too can only be 0 or 1. But it can be 0 while the first digit is 0 and 1, or it can be 1 while the 1st digit is 0 and 1. So we can count to four values with two digits. Adding a third digit position whose value is 0 or 1 allows a count to eight values. More and more digits provide more and more values. We illustrate the method in *Figure 1-9*. The decimal number is shown on the left with its binary equivalent in the center. The 4-bit binary number shown can represent 16 values. Two other number systems, *hexadecimal* and *binary coded decimal,* which we will see shortly, are on the right. Notice that the *least* significant bit (LSB) is in the 2^0 or 1's place. If a 1 appears in this place, a 1 is added to the binary count. The second place over from the right is the 2^1 or 2's place. A 1 appearing in this place means that 2 is added to the binary count. The weighted position value increases as digits are added from right to left. Just as in the decimal system, the value of a binary number is then the sum of the digits after each has been multiplied by its weight.

Decimal (XX_{10})	Binary (XXXX_2)	Hexidecimal (X_{16})	8421 BCD (XXXX_{BCD})	
0	0000 ← LSD	0		0000
1	0001	1		0001
2	0010	2		0010
3	0011	3		0011
4	0100	4		0100
5	0101	5		0101
6	0110	6		0110
7	0111	7		0111
8	1000	8		1000
9	1001	9		1001
10	1010	A	0001	0000
11	1011	B	0001	0001
12	1100	C	0001	0010
13	1101	D	0001	0011
14	1110	E	0001	0100
15	1111	F	0001	0101

Figure 1-9. Counting in the binary system with binary, hexadecimal and BCD.

Figure 1-10 shows a comparison between the decimal system and the binary system. The digit position values are 10 times the position value to the right in the decimal system (base-10), while the digit position values are two times the position value to the right in the binary system. Note that in both systems, the value of the number is obtained by adding together all the digit values after the digit has been multiplied by its weighted position value. Note also that a decimal number could be identified with a 10 subscript, while a binary number has a 2 subscript and a hexadecimal number has a 16 subscript. We normally do not write the 10 subscript with decimal numbers because they are so common that it is understood.

The following example will illustrate the principle of finding the decimal value of a binary number.

Example 6. Finding the Decimal Value of a Binary Number

What is the decimal equivalent of the binary number 101110_2?

Using the same procedure as for the decimal number in *Example 1-5,* each binary bit in the binary number produces a decimal equivalent for that binary place value. The *most significant bit* (MSB) of the binary number has a 1 or $1 \times 2^5 = 32$ is added to the value. The next significant bit has a 0 in its place, so it adds $0 \times 2^4 = 0$ to the value. The next three binary places have 1's, so they add $1 \times 2^3 + 1 \times 2^2 + 1 \times 2^1 = 8 + 4 + 2$. The *least significant bit* (LSB) is a 0 so it adds 0×2^0 to the value. The decimal value is, therefore, $32 + 0 + 8 + 4 + 2 + 0 = 46$.

It is also important to know how to convert from a decimal number to the equivalent binary number. A method of converting from decimal to binary is to divide repeatedly the decimal number by 2. The remainder generated by each division produces the binary number with the first remainder being the LSB. We illustrate in Example 7.

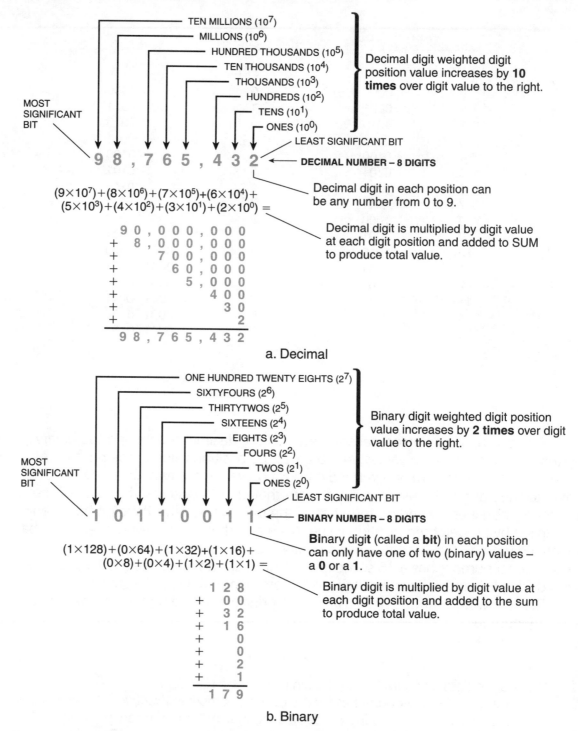

a. Decimal

b. Binary

Figure 1-10. Decimal and binary numbering systems.

Example 7. Converting Decimal to Binary

Convert 87_{10} to its binary equivalent.

$87_{10} / 2 = 43$ remainder of **1**^{LSB}
$43 / 2 = 21$ remainder of **1**
$21 / 2 = 10$ remainder of **1**
$10 / 2 = 5$ remainder of **0**
$5 / 2 = 2$ remainder of **1**
$2 / 2 = 1$ remainder of **0**
$1 / 2 = 0$ remainder of **1**^{MSB}
87_{10} = 1010111_2

Hexadecimal Numbers

Hexadecimal (called hex for short) is a popular number system used in digital electronics, especially in microcomputers. Each hex number is formed by grouping the binary bits in fours, starting from the right. This results in combinations from 0000_2 to 1111_2, or sixteen possible combinations. The first ten are represented by their decimal equivalents and the remaining six by the first six letters of the English alphabet; A, B, C, D, E, and F. Here is an example of a hex number converted to binary and decimal and vice versa.

Example 8. Converting Hex to Decimal and Decimal to Hex

Convert the hex number $3C_{16}$ to its binary and decimal equivalent. Use *Figures 1-9* and *1-10* to help you.

$3C_{16} = 0011\ 1100 = 00111100$
$3C_{16} = 111100 = 32 + 16 + 8 + 4 + 0 + 0 = 60$

The hex number may also be evaluated as a base-16 number:

$3C_{16} = 3 \times 16^1 + 12 \times 16^0 = 48 + 12 = 60$

To convert from binary to hex, reverse the process:

$11010101_2 = 1101\ 0101 = D5_{16}$

To convert from decimal to hex, apply the divide-by-16 method to convert 650 to its hex equivalent:

$650_{10} / 16 = 40$ with a remainder of $10^{LSD} = A_{16}$
$\quad 40 / 16 = 2$ with a remainder of $8 = 8$
$\quad\quad 2 / 16 = 0$ with a remainder of $2^{MSD} = 2$
$\quad\quad 650_{10} = 28A_{16}$

Its binary equivalent is:

$650_{10} = 0010\ 1000\ 1010$

Binary-Coded Decimal

The binary-coded decimal (BCD) code provides a way for decimal numbers to be encoded in a type of binary that is easily converted back to decimal. Each digit in the decimal number is assigned its equivalent binary code. Since digits from 0-9 need to be represented, a 4-bit code is required. This code is a weighted code more precisely known as 8421 BCD code. The 8421 part of the name gives the weight of each place in the 4-bit code. There are several other BCD codes that have other weights for the four places. The 8421 BCD, however, is the most popular so it is customary to refer to it as simply the BCD code. This code, shown in *Figure 1-9*, is useful for providing output to displays that are always numeric (0 to 9), such as those in digital multimeters, thermometers, and clocks. Converting decimal numbers to BCD and vise versa is quite simple as we shall see in the next example. To form a BCD number, simply convert each decimal digit to its 4-bit binary code. To convert BCD to decimal, just reverse the process. Note that the 4-bit code represents only numbers from 0 to 9, not the full possible sixteen numbers. It, therefore, is not as efficient as pure binary, but can save circuitry in certain applications.

Example 9. Converting Decimal to BCD and BCD to Decimal

Convert 369_{10} to BCD

$$369_{10} = \underset{0011}{\overset{3}{}} \quad \underset{0110}{\overset{6}{}} \quad \underset{1001_{BCD}}{\overset{9}{}}$$

Convert $0101\ 1000\ 0111_{BCD}$ to decimal

$$\underset{5}{\overset{0101}{}} \quad \underset{8}{\overset{1000}{}} \quad \underset{7}{\overset{0111}{}} = 587_{10}$$

Communications Codes — Bits, Nibbles, and Bytes

Communications transmission codes are made up of binary digits. (Recall that "bits" is the abbreviated form for "binary digits", and that each bit can have one of two possible states (high and low, on or off, 0 or 1). Four-bits combined in a uniform group is called a *nibble,* and an 8-bit group is called a *byte.* Many modern computers operate with a 16-bit or 32-bit word size, so each word comprises two bytes or four bytes, respectively.

A code is a previously agreed upon set of meanings that defines a given set of symbols and characters. Digital systems process only codes consisting of 0's and 1's (binary codes). You have learned that coded patterns can represent operations on the TV remote control, and that by a telegraph system you can send a message of characters (letters, numbers or symbols). It was virtually impossible to automate the telegraph because of the varying number of dots and dashes for the various characters. What was needed was a code that had an equal number of equal duration signaling elements for each character. A standardized alphanumeric code called the American Standard Code for Information Interchange (ASCII) is the most widely used in industry.

ASCII

Figure 1-11 shows the ASCII characters and their associated codes. Compare this chart to the Morse code in *Figure 1-8.* Not only does the Morse code have a varying number of elements (dots and dashes) for each character, it is also restricted to the letters, numbers, and a few punctuation marks. ASCII has not only both upper and lower case for letters but also has a number of other features that make it very versatile. For example, the ASCII format is arranged so that lower case letters can be changed to upper case by changing only one bit of the seven. By changing bit six from a 0 to a 1, an upper case letter is changed into lower case. Also, bits 4 through 1 of the numeric characters 0-9 are the binary-coded-decimal (BCD) value of the number. ASCII is the most commonly used code for the input and output of computers. The keystrokes on the keyboard of a computer are converted directly into ASCII for processing by the digital circuits. In Example 10, we determine the codes that are entered into a computer's memory when a line of program is typed on its keyboard.

Signals and Switches

We pointed out in the beginning of this chapter that, while many quantities are inherently analog in nature, digital quantities also exist. In fact, virtually any analog quantity can also be represented digitally. Analog values are frequently converted to a digital value for more convenient processing and display purposes. An example is the gasoline pump where the smooth analog flow of gasoline is measured and the volume is displayed digitally to the nearest tenth of a gallon or liter.

Bit Position

| | | | | | | | 0 | 1 | 0 | 1 | 1 | 0 | 0 | 1 |
| | | | | | | | 0 | 0 | 1 | 1 | 1 | 1 | 0 | 0 |
1	2	3	4	5	6	7	1	1	1	1	0	0	0	0
0	0	0	0				@	P	'	p	0	sp	NUL	DLE
1	0	0	0				A	Q	a	q	1	!	SOH	DC1
0	1	0	0				B	R	b	r	2	"	STX	DC2
1	1	0	0				C	S	c	s	3	#	ETX	DC3
0	0	1	0				D	T	d	t	4	$	EOT	DC4
1	0	1	0				E	U	e	u	5	%	ENQ	NAK
0	1	1	0				F	V	f	v	6	&	ACK	SYN
1	1	1	0				G	W	g	w	7	'	BEL	ETB
0	0	0	1				H	X	h	x	8	(	BS	CAN
1	0	0	1				I	Y	i	y	9	)	HT	EM
0	1	0	1				J	Z	j	z	:	*	LF	SUB
1	1	0	1				K	[	k	{	;	+	VT	ESC
0	0	1	1				L	\	l			<	,	FF	FS
1	0	1	1				M	]	m	}	=	-	CR	GS
0	1	1	1				N	^	n	~	>	.	SO	RS
1	1	1	1				O	_	o	DEL	?	/	SI	US

Figure 1-11. ASCII characters and their associated codes.

Example 10. How ASCII Codes Represent Information

Encode the following line of program into ASCII using *Figure 1-11*. Remember bit position 1 is the LSB.

30 PRINT "C=";y

The solution is found by looking up each character in *Figure 1-11*.

Character	ASCII	Hexadecimal
3	0110011	33_{16}
0	0110000	30_{16}
Space	0100000	20_{16}
P	1010000	50_{16}
R	1010010	52_{16}
I	1001001	49_{16}
N	1001110	$4E_{16}$
T	1010100	54_{16}
Space	0100000	20_{16}
"	0100010	22_{16}
C	1000011	43_{16}
=	0111101	$3D_{16}$
"	0100010	22_{16}
;	0111011	$3B_{16}$
y	1111001	79_{16}

Analog and digital signals are used to represent all types of data, including physical data. Data is any kind of fundamental element of information, such as quantities, numbers values, letters, words, and symbols capable of being processed. We have considered several examples of how analog and digital signals represent data in this chapter. Both analog and digital signals can represent time, temperature, sound recording. We have seen a multimeter that can display electrical current, voltage, and resistance. However, we have not discussed in any detail how these two states even allow us to build any useful electronic circuits to do even simple processing or problem solving. So let's examine how we can use simple switches to represent numbers and provide the basis for digital electronics and even a digital computer.

Digital States Against Time — Timing Diagrams

What modern convenience comes to mind when you hear the terms "on" and "off"? A light switch, maybe? This simple example of a common digital circuit is illustrated in *Figure 1-12*. It has two states, no in-between, just light on and light off. When the switch is open, the light is off. When the switch is closed, the light is on. In digital circuits, the two voltage levels, the two current levels, or the two light conditions represent the two binary digits, 0 and 1. *Figure 1-12d* shows the current in the circuit or the voltage across the lamp plotted against time. It is called a timing diagram. Timing diagrams are used to show the "high" and "low" levels of a digital signal as it changes relative to time. The vertical axis of the plot displays the voltage or current level, and the horizontal axis displays the time. We will discuss the importance of timing in digital circuits in the next chapter.

Light-Emitting Diode Digital Circuit

Any type of lamp may be used to represent digital lows and highs (0 and 1), but a light-emitting diode (LED) is used extensively in digital circuits to display the "on " and "off" conditions of the digital signal. A typical LED and its circuit diagram symbol are shown in *Figure 1-13*. The LED is connected to a single-pole single-throw switch, a current limiting resistor, and a 5-volt power supply. When the switch is open, there is no current and the LED is "OFF." When the switch is closed, there is current and the LED is "ON."

The advantage of the LED over a incandescent lamp is its low power consumption. An LED can produce a visible amount of light with only a few milliamperes of current, so it produces very little heat. Another advantage is its very long life compared to a lamp.

The most common type of LED is the one that produces red light, though other colors are available. The LEDs actually produce different light colors, not just different color lens caps as with incandescent lamps. Unlike the incandescent lamps, the LED is a directional device; that is, it has to be connected in the circuit with a forward bias to produce light. When it is forward-biased by approximately two volts, it starts conducting current and produces light. Forward-bias is when the anode is positive in voltage relative to the cathode. When the LED is reversed-biased (negative voltage on its anode), it has zero current and gives off no light.

Groups of LEDs for Displaying Binary Numbers

Groups of four LEDs can be used to represent binary numbers that can be read as hexadecimal numbers. The conditions are: when an LED is ON, it represents a HIGH or 1 condition; when it is off, it represents a LOW or 0. The LEDs in *Figure 1-14* represent two hex numbers that can be changed by their corresponding switches. Example 11 shows how the switches can be used to represent (encode) hex numbers or ASCII characters.

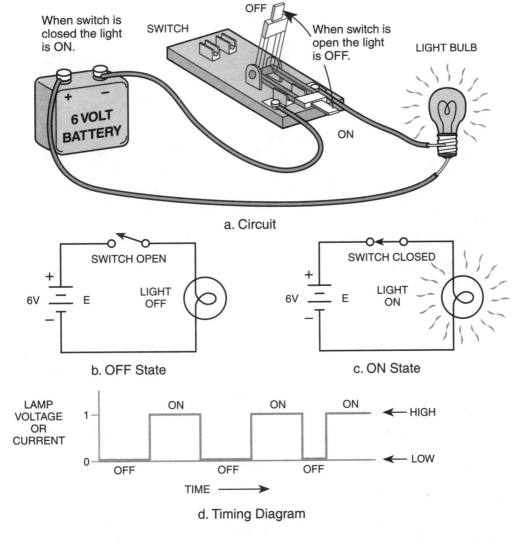

a. Circuit

b. OFF State

c. ON State

d. Timing Diagram

Figure 1-12. A light switch example of a digital circuit with its waveforms and timing diagrams.

Example 11. How LEDs Represent Binary Numbers

A: What are the hexadecimal numbers represented by the LEDs if the switches in *Figure 1-14* are in the following states? Use *Figure 1-9* to help you.

MSBON-OFF-ON-ON   OFF-ON-OFF-OFFLSB

   1   0   1   1   0   1   0   0

       B   4

B: What ASCII character is represented by the seven bits not counting the MSB if the switches are in the following states? Use the ASCII table in *Figure 1-11* to help you. Remember bit 7 is the one next to the MSB, and bit 1 is the LSB.

MSBON-ON-OFF-OFF   OFF-ON-ON-OFFLSB

   1   1   0   0   0   1   1   0

           F

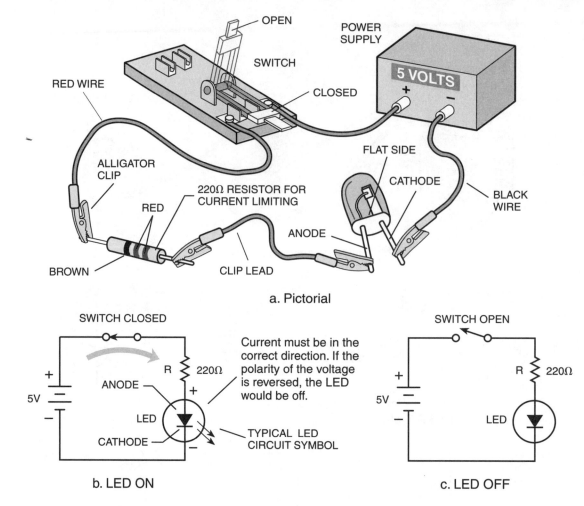

a. Pictorial

b. LED ON

SWITCH CLOSED

Current must be in the correct direction. If the polarity of the voltage is reversed, the LED would be off.

R ≥ 220Ω

5V

ANODE

LED

CATHODE

TYPICAL LED CIRCUIT SYMBOL

c. LED OFF

SWITCH OPEN

R ≥ 220Ω

5V

LED

Figure 1-13. An LED in a typical digital circuit.

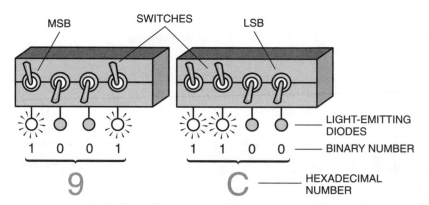

Figure 1-14. Switches control LEDs to represent hex numbers in binary code.

Digital Integrated Circuits

Most of the circuits in digital equipment are in the form of integrated circuits (ICs). This means that all of the circuit is fabricated on a single silicon chip; that is, all of the components, such as transistors, resistors and diodes are fully formed and interconnected on a single silicon chip. The components are very tiny, so many digital circuits can be formed on a very small chip. For example, the microprocessors in personal computers have digital integrated circuits with a million or more transistors on a silicon chip as small as a fingernail. The primary advantages of digital integrated circuits are small size, low cost, and outstanding reliability. We will investigate integrated circuits in more detail in Chapter 7. Now, in the next chapter, let's look at the important functions that are required in digital electronic systems.

Quiz for Chapter 1

1. Quantities that vary at a continuous rate are called
   a. linear
   b. digital
   c. analog
   d. logical
2. Electronic circuits can process information about physical quantities easier and more accurately with
   a. linear circuits
   b. digital circuits
   c. analog circuits
   d. logic circuits
3. An example of a digital audio source is
   a. AM radio
   b. compact disc
   c. phonograph
   d. standard cassette tape
4. If the amplification of the audio amplifier is 50 and the signal voltage output of the radio detector is 0.2 volt, at one-fourth volume control settings, the power delivered to an 8-ohm speaker is
   a. 1.25 watts
   b. 12.5 watts
   c. 0.78 watts
   d. 2.5 watts
5. For the clock in *Figure 1-6;* What time is it if the crystal oscillator has put out 52,200 one second pulses? The clock starts at 12:00.
   a. 10:20 AM
   b. 2:30 PM
   c. 4: 45 PM
   d. 10:20 PM
6. A TV remote control is an example of
   a. digital coding
   b. analog coding
   c. hexadecimal numbers
   d. Morse code
7. A two-state or two-valued system is referred to as a
   a. analog system
   b. digital system
   c. BCD system
   d. binary system
8. Decode this message from the International Morse Code:

   —•• •• — —•• •• — •— •—••

   a. DECIMAL
   b. DECODE
   c. DECIBEL
   d. DIGITAL

9. The value of a number is the sum of the digit values after each digit has been multiplied by its
   a. weight
   b. power
   c. radix
   d. base
10. Convert 63_{10} to its binary equivalent.
    a. 0000111111
    b. 0001010101
    c. 0010110110
    d. 0011100011
11. What is the decimal equivalent of the binary number 101010_2?
    a. 24
    b. 40
    c. 42
    d. 124
12. Convert the hex number $B5_{16}$ to its binary equivalent.
    a. 1100 0011
    b. 1000 0111
    c. 0101 1011
    d. 1011 0101
13. Convert 0011 1001 0111_{BCD} to decimal.
    a. 468_{10}
    b. 284_{10}
    c. 953_{10}
    d. 397_{10}
14. Decode the ASCII characters below using the table of *Figure 1-11*.
    1010010 1010101 1001110 0100001
    a. RUN!
    b. JUMP
    c. WHO?
    d. RAMP
15. When positive polarity is applied to an LED's anode compared to the cathode, it is
    a. "OFF"
    b. Reversed biased
    c. Forward-biased
    d. A high impedance

Questions and Problems for Chapter 1

1. For the circuit in *Figure 1-4a,* if the amplifier's gain is 50, the speaker is 8 ohms, and the detector's output is 50 mV, what is the power delivered to the speaker with the volume control set at full, one-half, and one-tenth volume?

2. a. For the circuit in *Figure 1-5b,* what is the current in the motor if the speed control is set to one-fourth the resistance from A to B?

   b. Is this faster or slower than with the control set to one-half?

3. For the digital clock in *Figure 1-6,* if it starts at 12:00 noon, what time will it be after the crystal oscillator has put out 25,400 one-second pulses?

4. For the digital clock in *Figure 1-6,* if it starts at 12:00 noon, what time will it be after the crystal oscillator has put out:

   a. 3600 one-second pulses?

   b. 46,800 one-second pulses?

5. a. Is a binary system digital?

   b. Is a digital system binary?

6. Separate the digits of the decimal number 3572 into their weighted digit positions.

7. What is the decimal equivalent of the binary number 100101_2?

8. What is the binary equivalent of the decimal number 63?

9. Convert the hex number $5C_{16}$ to its: a. binary equivalent b. decimal equivalent?

10. Convert the decimal number 47 to its: a. hex equivalent b. binary equivalent?

11. Encode the following line of code into ASCII using *Figure 1-11:*

    DIGITAL ELECTRONICS

12. Decode the following ASCII message from its hex code:

    41, 4E, 41, 4C, 4F, 47, 20, 56, 4F, 4C, 54, 41, 47, 45

CHAPTER 2

Digital System Functions

Looking Back

In Chapter 1, the examples of common analog and digital systems showed the difference between the two. Recall that analog system values vary continuously, but digital system values are comprised of discrete parts formed into codes to represent the values. The digital system uses these codes that change with time to carry information from one point to another through the system. The codes are processed to accomplish particular functions that combine to make up the complete digital system.

This chapter provides an overview of basic functions that make up digital systems. This overview will help you understand how a digital system works and accomplishes tasks. Only basic concepts will be covered here; more details will be presented in later chapters.

Returning to Codes

Chapter 1 showed different types of digital codes, including Morse code and ASCII, for representing alphabetic characters, numbers, punctuation marks, special characters, and commands. Each of the codes uses *two* values (*two* levels, ON-OFF, 0 and 1) for each of the bits (bit is a contraction of **bi**nary dig**it**) in the *binary* code. Recall that the bit is the smallest unit of binary information, and since it is binary, it can have only *two* values. How then do we represent 10 different values?

Figure 2-1 will help explain. A 2-bit code is required to represent four values, a 3-bit code for eight values, and a 4-bit code for 10 values, and so on. But note that a 4-bit code can represent not just 10 values, but 16 values. To represent more and more values (or different things), we just add more bits to the code. For a base-2 (binary) system, the number of bits n required to distinguish K different things is given by

$$2^n = K$$

Note that, even though the codes are binary numbers, they can represent whatever the system designer designates them to be; e.g., the control system commands shown. In ASCII, a different binary code (binary number) is used for each of the characters.

If the digital code can represent more values than required, the coding efficiency (in percent) is:

$$\text{Percent Coding Efficiency} = \frac{\text{Values required}}{2^N} \times 100$$

Two Values	
0	First value
1	Second value

a. 1-Bit Code

Four Values	
0 0	First value
0 1	Second value
1 0	Third value
1 1	Fourth value

b. 2-Bit Code

Eight Values	
0 0 0	First value
0 0 1	Second value
0 1 0	Third value
0 1 1	Fourth value
1 0 0	Fifth value
1 0 1	Sixth value
1 1 0	Seventh value
1 1 1	Eighth value

c. 3-Bit Code

Ten Values		Control System Commands
0 0 0 0	First value	0. System Idle
0 0 0 1	Second value	1. Master Sw ON
0 0 1 0	Third value	2. Close Output Chute
0 0 1 1	Fourth value	3. Open Input Chute
0 1 0 0	Fifth value	4. Weigh Input Material
0 1 0 1	Sixth value	5. Close Input Chute
0 1 1 0	Seventh value	6. Start Conveyor Belt
0 1 1 1	Eighth value	7. Open Output Chute
1 0 0 0	Ninth value	8. Stop Conveyor Belt
1 0 0 1	Tenth value	9. Master Sw OFF
1 0 1 0	Eleventh value	Available for Additional Commands
1 0 1 1	Twelfth value	
1 1 0 0	Thirteenth value	
1 1 0 1	Fourteenth value	
1 1 1 0	Fifteenth value	
1 1 1 1	Sixteenth value	

d. 4-Bit Code

Figure 2-1. Bits are added to binary codes to represent more and more binary numbers, values, or commands.

Example 1. Determining Number of Bits to Represent Objects

How many bits are required to represent 22 different objects?

We can get the exact answer with logarithms, but it may be simpler by trying powers of 2. If we try $2^3 = 8$ and $2^4 = 16$, neither is enough. Trying $2^5 = 32$ is more than enough, but since a fraction of a bit cannot be implemented in practice, we must use the next highest whole number; that is, 5 bits. So, for a base-2 system, the number of bits n is 5 for $K = 22$ objects. Since $2^5 = 32$, there are 10 values included in the 5-bit code that are not used in this application. (Remember that the 32 values are 0 through 31, inclusive. The values 0 through 21 are used, so the unused values are 22 through 31.)

Transmission of Digital Information

Every digital system requires the movement of digital information from one point in the system to another. It may be called code transfer, binary bus transfer, digital communications, or data communications. Whatever it's called, it means that a binary code representing information is moved from one place to another. For want of a better term, we will use *data communications* because information termed "data" refers to numerical, alphabetical, or special-purpose characters which represent some message or intelligence. Regardless of the origin of the data, a binary digital signal — a code — consists of a sequence of bits, appropriately grouped.

Movement of data requires a *data communications system*. The three components that make up a data communications system are: the source or transmitter, the channel or transmission path, and the receiver. Electrically, in a digital system, the bit value may be represented by a voltage or current value.

Data are commonly transferred between the transmitter and the receiver, as shown in *Figure 2-2,* by changes in current or voltage on the channel or transmission line(s) between the units. For example, to transmit an ASCII character, a "1" or MARK could be represented by a positive voltage or current pulse; a "0" or SPACE could be represented by a zero voltage or no current condition. The data — a code of 0100101, one digit at a time in sequence — is being transmitted from one point to another in *Figure 2-2.* Notice that a START bit is included just before the character code and a STOP bit is included immediately after the character code. The purpose of these bits is probably obvious, but they will be explained later.

In digital systems, two methods — parallel and serial — are used to move binary information from transmitter to receiver. Let's look at these two types of data transmission and the characteristics and advantages of each.

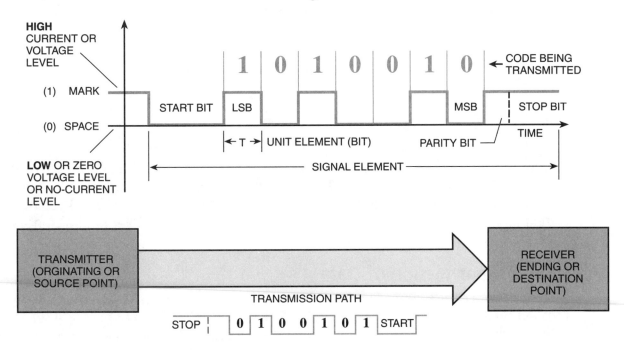

Figure 2-2. Data Communications System — Data transmitted from one point to another by voltage or current pulses.

Parallel Transmission of Data

Parallel transmission is often used for short distance communications between digital equipment. Parallel transmission moves the data characters using multiple lines — one for each bit. The 1 or 0 on each data line representing a bit of the digital code appears on all of the multiple lines at the same time. In *Figure 2-3,* we are transmitting one bit of a 7-bit ASCII character on each line of the seven lines. Each bit of a character travels on its own line. As shown in *Figure 2-3,* each of the 7-bit ASCII characters appears in sequence as time proceeds. Thus, the transmission is also known as *parallel-by-bit, serial-by-character* transmission. It is often used over short distances, such as from a computer to a printer, or from one circuit to another inside a piece of digital equipment. Parallel transmission is much faster, so it is employed

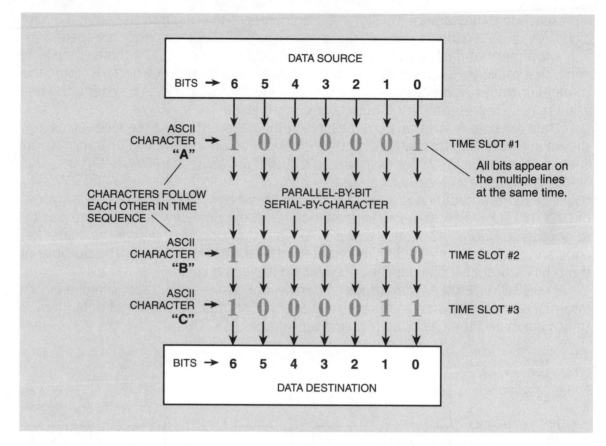

Figure 2-3. Parallel-by-bit, serial-by-character transmission of data.

when the equipment being connected is close by (usually less than 25 feet). As the distance between equipment increases, the multiple lines not only become more costly, but the difficulty of transmitting pulse signals over long lines and keeping them in correct time relationship becomes a major limitation. Over long distances, serial transmission is preferred because it involves only one transmission line.

Serial Transmission of Data

To send data over a single line, the bits must be transmitted in series (appropriately called serial transmission). The 7-bit ASCII character "C" would appear in a series time sequence as shown in *Figure 2-4*. Both the transmitter and the receiver use the same nominal unit element time (called clock rate) shown in *Figure 2-2*, so the system must provide a way for the receiver to duplicate the transmitter clock so it can decode the data correctly. The communications system must ensure that the receiver clock is timed so that errors do not occur. If the characters are seven bits long, the receiving device is designed to know which of the seven bits it is supposed to look for at any one instant.

You might ask: "How does the receiver know when one character ends and the next one begins. How does the equipment get *synchronized*?"

Timing Is Very Important

We will get back to the data communications example shortly, but first let's define synchronization. Synchronization (or sync for short) means "To do things at the *same time*" and "To cause to occur or operate with *exact coincidence in time or rate*." Since digital systems depend on digital signals which are changing from 0 to 1 or 1 to 0 levels as the system goes through its tasks, it is very important that the system has accurate synchronization or accurate timing.

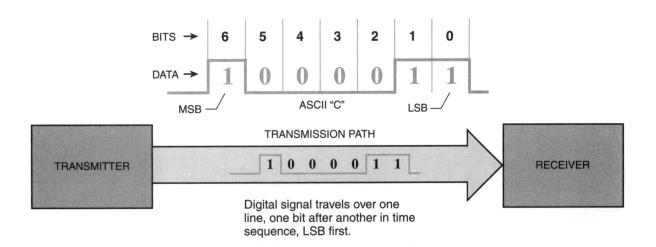

Figure 2-4. Serial Transmission — Bits transmitted over a single line in series time sequence; in this case, an ASCII code representing the letter "C".

Synchronization

Let's demonstrate synchronization using *Figure 2-5.* A timing generator, called a *clock*, generates regular pulses, called *clock pulses*, that are very accurate in frequency; therefore, the *time interval* when the clock pulses are at the 1 level and at the 0 level are very accurately controlled. In addition, the *transitions* from 1 to 0 and 0 to 1 occur at very accurate times, and are used to time or "trigger" action in digital circuits throughout the digital system. For example, in *Figure 2-5,* the transition from 1 to 0 of the clock pulse triggers a sampling circuit that samples the bit of the received data in the middle of the bit time. This allows a variation in the received data of one-half the bit time before an error occurs. We will see many applications of clock pulses throughout this book as we discuss digital electronic circuits.

Synchronizing Transmitter and Receiver

Now, back to the data communications system. We can use it to demonstrate synchronization. First of all, both transmitter and receiver must be operating at the same clock rate. Thus, in serial transmission, you will hear that the digital system is operating at the same *baud rate.* That means the unit element time shown in *Figure 2-5* is the same. In addition to synchronizing its clock with the transmitter clock, the receiving device must be able to determine when to start counting an individual character so that the individual bits can be sampled and identified. Unique marker characters or bits are required to indicate the beginning and end of a character or group of characters. *Figure 2-2* shows start and stop bits, and *Figure 2-5* shows the sampling. In parallel transmission, a signal called the *strobe* or clock on an additional line indicates to the receiver when all the bits are present so they can be read or sampled.

In *synchronous systems* for high-speed data communications, data is transferred in blocks which are framed by STX (start-of-text) and ETX (end-of-text) characters, as shown in *Figure 2-6.* The receiving device synchronizes its clock when it receives an agreed upon SYN character. After the series of SYN characters, the actual data is preceded by an STX character and followed by an ETX character. These two characters tell the receiver when the actual data begins and ends.

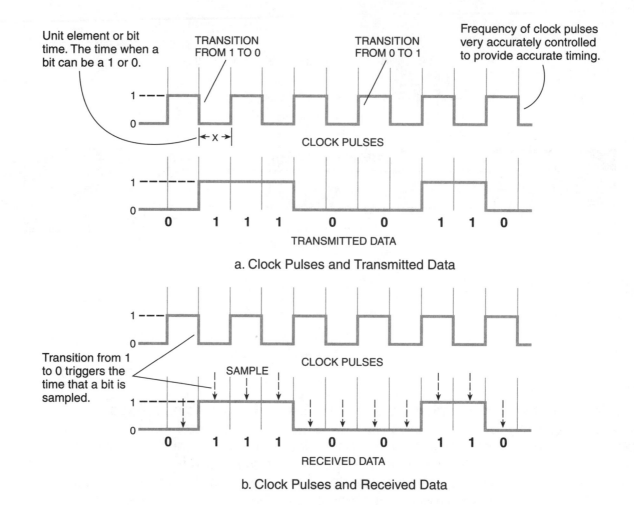

a. Clock Pulses and Transmitted Data

b. Clock Pulses and Received Data

Figure 2-5. Synchronization is provided by a clock generator controlled to a very accurate frequency. Bit times and transition times are held very accurately.

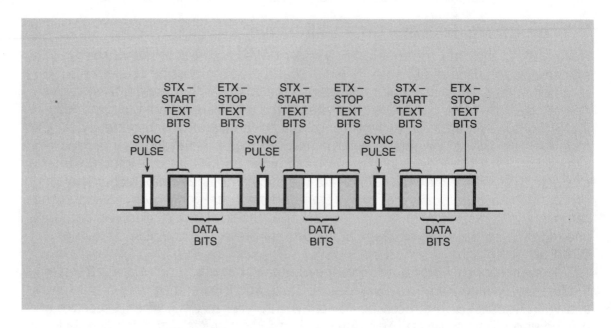

Figure 2-6. The "start text" bits tell the receiver that characters are coming and "stop text" bits tell it that the characters have ended. Sync pulses synchronize transmitter and receiver.